China Water Saving Irrigation

Rural Water Resources Department, Ministry of Water Resources, P. R. China
China Irrigation and Drainage Development Center

China WaterPower Press
www.waterpub.com.cn

图书在版编目（CIP）数据

中国节水灌溉 = China Water Saving Irrigation：英文 / 中华人民共和国水利部农村水利司，中国灌溉排水发展中心编. -- 北京：中国水利水电出版社，2010.12
ISBN 978-7-5084-8274-3

Ⅰ. ①中… Ⅱ. ①中… ②中… Ⅲ. ①农田灌溉－节约用水－中国－英文 Ⅳ. ①S275

中国版本图书馆CIP数据核字(2010)第264204号

审图号：GS（2009）1210号

书　　名	**China Water Saving Irrigation**
作　　者	Rural Water Resources Department, Ministry of Water Resources, P. R. China China Irrigation and Drainage Development Center
出版发行	中国水利水电出版社 （北京市海淀区玉渊潭南路1号D座　100038） 网址：www.waterpub.com.cn E-mail：sales@waterpub.com.cn 电话：（010）68367658（营销中心）
经　　售	北京科水图书销售中心（零售） 电话：（010）88383994、63202643 全国各地新华书店和相关出版物销售网点
排　　版	中国水利水电出版社装帧出版部
印　　刷	北京鑫丰华彩印有限公司
规　　格	210mm×285mm　16开本　4.25印张　168千字
版　　次	2010年12月第1版　2010年12月第1次印刷
印　　数	0001—1000册
定　　价	38.00元

凡购买我社图书，如有缺页、倒页、脱页的，本社营销中心负责调换

版权所有·侵权必究

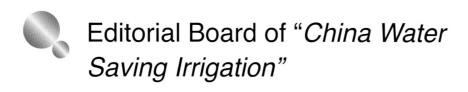

Editorial Board of "*China Water Saving Irrigation*"

Chairman: E Jingping

Vice Chairmen: Wang Xiaodong, Li Yangbin

Members: Ni Wenjin, Gu Binjie, Wang Xiaoling, Wu Yuqin

Chief Editors: Wang Xiaodong, Li Yangbin

Deputy Chief Editors: Ni Wenjin, Gu Binjie, Wang Xiaoling

Editors: Wu Yuqin, Feng Guangzhi, Ren Xiaoli, Li Yingneng, Guo Huibin, Pan Yunsheng, Liu Yunbo, Zhang Yuxin, Gu Tao, Long Haiyou, Li Tienan, Wang Yue, Bai Jing, Sun Dongxuan

Translation Editor: Mu Jianxin

Translators: Gu Tao, Wang Yue, Long Haiyou

PREFACE

Water is not only an irreplaceable basic element in the process of every life, but also a vital foundation resource to maintain the national economic and social development. Affected by the natural and geographical conditions as well as the population, China is a large agricultural but water deficient country. Besides, the development of agriculture has strong dependency on irrigation. The CPC Central Committee and the State Council attach great importance to the development of water saving irrigation, it has been requested by the Third Plenary Session of the Fifteenth CPC Central Committee to extend water saving irrigation as a revolutionary measure, to increase water use efficiency substantially and to make great efforts to expand command irrigation area. By the end of 2008, China's command irrigation area has increased to 0.877 billion mu[1], including 0.367 billion mu of water-saving irrigation area. It has produced 75% of grain and 90% of cash crops of the national total, on the basis of the irrigated area accounting for 48% of the national cultivated land. In the 30 years since the reform and opening-up, the command irrigation area has increased by 0.12 billion mu, the grain production has increased by 50%, and the irrigation water use efficiency has increased from about 0.35 to 0.48 while the total irrigation water use has kept unchanged. It should be concluded as a result that the development of irrigation has made significant contributions in guaranteeing the national food security and in increasing farmers' income, and the development of water saving irrigation has played an important role in promoting the sustainable water resources use and the sustainable socio-economic development.

Since the 1980s, China's water saving irrigation has entered into a stage of rapid development. Accompanied by implementation of a number of key water saving irrigation schemes, such as rehabilitation of large and medium irrigation districts, the water saving schemes in key counties to increase agricultural production, the water saving irrigation demonstration plot, the construction of small-scale irrigation and drainage schemes, and so on. The water saving irrigation technology system and policy system have been gradually formed a line with China's actual conditions. A number of mature technologies have been researched, developed, reserved and extended. The water saving irrigation technology modes appropriate to different regions have been summed up. A set of service system for the extension of water saving irrigation techniques has been formed tentatively. All of these have laid a solid foundation for the development of water saving irrigation in the future.

With the rapid economic and social development of China, water shortage will become more

---

[1] Mu is equal to 1/15 hectare.

prominent. As the big water consumer in China's economy and society, agricultural irrigation is the sector with great potentials for water savings, and also the main battlefield for the development of water saving society, on the assumption that there still exists a huge gap in water use efficiency compared to the developed countries. It has been required by the Third Plenary Session of the 17th CPC Central Committee to basically complete the rehabilitation of large irrigation districts by 2020. According to the requirements of relevant planning, by 2030, the ratio of water saving irrigation area to command irrigation area should reach over 80%, the irrigation water use per mu in normal year should be controlled with in 390m^3, and the national average on-farm irrigation water use coefficient should be over 0.6 while the agricultural irrigation water use maintain unchanged on the whole.

At the same time, China's agricultural modernization is on the way with Chinese characteristics. Modern agricultural development has provided a higher demand on advanced irrigation methods and precision irrigation management. It becomes an inevitably requirement to greatly develop advanced water saving irrigation technologies, including sprinkler irrigation, micro-sprinkler irrigation and the others. Therefore, it is a vital and urgent task for us to continuously develop water saving irrigation as a revolutionary measure, to promote the development of new socialist countryside and water saving society, to guarantee the national food security, to continuously increase farmers' income, and to boost the sustainable economic and social development with the sustainable use of water resources.

"*China Water Saving Irrigation*" is compiled and published by Department of Irrigation, Drainage and Rural Water Supply and China Irrigation and Drainage Development Centre, Ministry of Water Resources, P.R.CHINA. This is a useful work as it comprehensively summed up and introduced the achievements, technologies, and policies of water saving irrigation. It will play active roles in promoting the development of water saving irrigation in China, and in disseminating and extending water saving irrigation technologies to the whole society.

October 2009

Contents

PREFACE

1 Water and Agriculture	1
1.1 Water and Drought	1
1.2 Irrigation and Agriculture	4
2 Overview of Water Saving Irrigation Development	11
2.1 Progress of Water Saving Irrigation Development	11
2.2 Scientific and Technical Support to the Development of Water Saving Irrigation	15
2.3 Water Saving Irrigation Facilities and Equipments	16
3 Strategy on Water Saving Irrigation Development	17
3.1 General Train of Thought	17
3.2 The Way Forward toward Water Saving Irrigation by Regions	18
4 Technical System and Index of Water Saving Irrigation	27
4.1 Technical System of Water Saving Irrigation	27
4.2 Technical Index of Water Saving Irrigation	28
5 Main Measures of Water Saving Irrigation	29
5.1 To Rationally Allocate and Efficiently Use Irrigation Water Sources	29
5.2 To Increase Water Conveyance and Allocation Efficiency	31
5.3 To Increase Water Use Efficiency of Surface Water Irrigation	38

5.4 Sprinkler Irrigation	41
5.5 Micro Irrigation	44
5.6 Dibble Seeding and Water Injecting Technology	45
5.7 Agronomic Water Saving Technologies and Measures	45
5.8 Management Water Saving Technologies and Measures	46
6 Laws, Regulations and Policies on Water Saving Irrigation	49
6.1 Laws and Regulations	49
6.2 Policies for Agricultural Water Conservation	50
6.3 Technical Standards on Water Saving Irrigation	52
7 Effects from Water Saving Irrigation	53
7.1 Increased Irrigation Water Use Efficiency	53
7.2 Increased Water Productivity and Overall Agricultural Production Capacity	55
7.3 Speeded up the Transformation from Traditional Agriculture to Modern Agriculture	56
7.4 Relieved the Contradictions between Water Supply and Water Demand, Improved the Ecological Environment in Some Areas	57
7.5 Promoted the Development of Water Saving Irrigation Equipment and Facility Industry	58
8 Conclusions	59

1 Water and Agriculture

1.1 Water and Drought

1.1.1 Precipitation

In China, the depth of average annual precipitation is 650mm with an amount of 6.1775 trillion m^3. Affected by monsoon climate and geographic features, the average annual precipitation decreases from the southeast to the northwest. The land area in the south of China accounts for 36% of the national total, but the precipitation occupies 68%. While the land area in the north of China holds 64% of the national total, however the precipitation only occupies 32%. The precipitation varies interannually, thus caused continious wet or drought. The allocation of precipitation during the year is extraordinarily uneven, the largest average precipitation in four consecutive months accounts for about 55% of the average annual total in the south of China. While in the north of China, the largest average precipitation in four consecutive months is more than 70% of the average annual total.

1.1.2 Water Resources

China's mean annual water resources are approximately 2.8412 trillion m^3 (excluding the duplication between surface water and groundwater), including an estimated 2.7388 trillion m^3 of surface water and 0.8218 trillion m^3 of groundwater. Although the total amount of water resources rank 6th in the world, the per capita water resources is only about 2100 m^3, which only accounts for 28% of the world average, making China one of the 13 water-deficient countries of the world. The land and water resources do not match in China. The water resources in the south of China accounts for 81% of the national total, but the arable land is only 40% of the national total and the population is 54% of the national total; while in the north of China, the water resources make up 19% of the national total, however the arable land accounts for 60% of the national total and the population accounts for 46% of the national total.

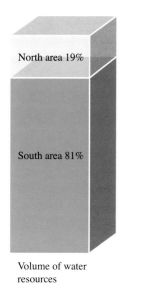
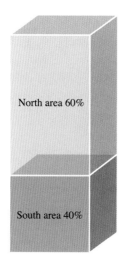

Matching Figure of Water and Soil Resources

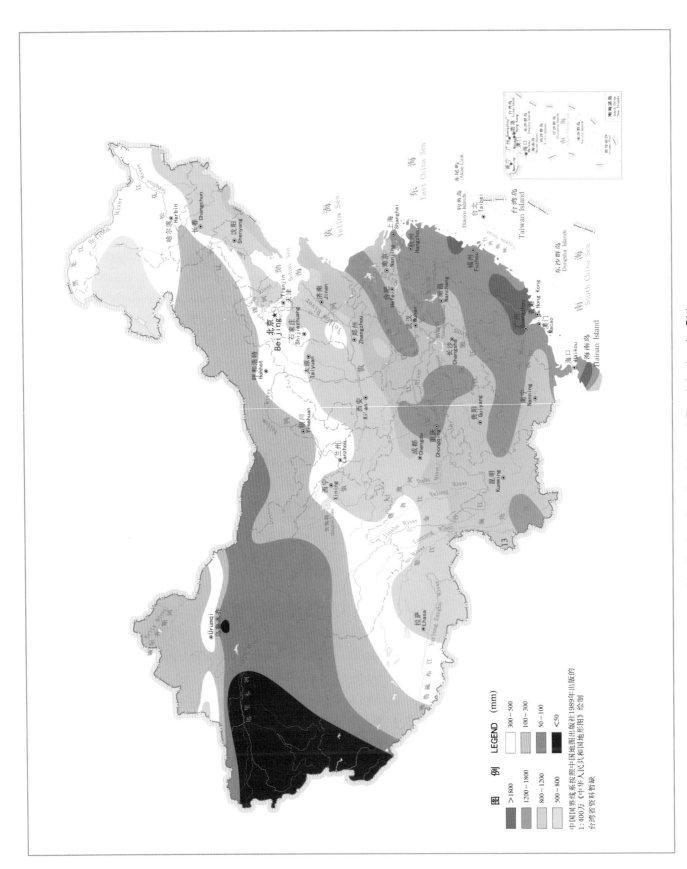

Distribution Graph of Annual Precipitation in China

1.1.3 Drought Disasters

China's drought disasters occur frequently. Drought in the spring is very serious to the north of the Qinling Mountain Divide and Huaihe River with a frequency of nine in ten years, sometimes these droughts are even accompanied with summer droughts or, summer to autumn droughts. Summer droughts or consecutive summer to autumn droughts are very common in the middle and lower reaches of the Yangtze River, dominated with summer droughts; consecutive spring to summer droughts are very common in the upper reach of the Yangtze River and the southwestern region, dominated with spring droughts; winter droughts and spring droughts are very common in South China, dominated with spring droughts, or with summer droughts or autumn droughts in some regions.

Drought disasters have caused huge losses to agricultural production. The losses of grain production due to drought disasters account for more than 50% of the total grain production losses caused by various natural disasters, and account for approximately 5% of the total amount of average annual grain production. According to the statistics, the average annual drought-stricken area in China was 24.9 million hectares in the 1990s, and was 25.7 million hectares on average since the 21st century. The year of 2000 was a serious dry year, the total losses from grain production as a result of drought disaster was 60 billion kg, accounting for 13% of total grain production, and the losses from cash crops reached up to RMB 51.1 billion Yuan.

1.2 Irrigation and Agriculture

1.2.1 Overview of Agriculture

China's arable land area was 121.7352 million hectares in 2008, less than half of the world average. The cultivated lands are mainly distributed in the Northeast Plain, the North China plain, the middle-lower Yangtze River plain, the Pearl River delta and Szechwan Basin et al.

Since the land use efficiency is very high, intensive cultivation is a tradition in China's agriculture. The total sown area of crops was 156.266 million hectares in 2008, out of which the ratio of the planting area of grain crops and cash crops was 2.14:1, and the cropping intensity was 1.28. Compared with 1949, in 2008 the total grain production, the average grain yield per unit area, and rural residents annual net income in China have increased by several to a hundred times.

In 2008, China's total agricultural output was worth 2.80442 trillion Yuan, the forestry output was 215.29 billion Yuan, the total animal husbandry output was 2.05836 trillion Yuan, and the total fishery output was worth 520.34 billion Yuan. The composition of output value was 13:1:9.6:2.4 among agriculture, forestry, animal husbandry and fishery.

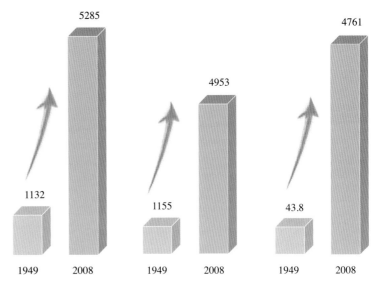

Grain total production (10^8kg)

Grain production per hectare (kg/hm^2)

Rural residents annual net income (Yuan)

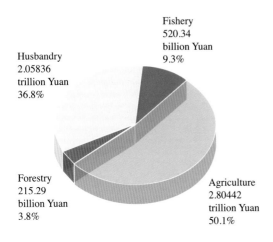

Proportion Chart of Agriculture, Forestry, Husbandry and Fishery

1.2.2 The Status of Irrigation

China can be divided into three different irrigation zones based on precipitation: perennial irrigation zone where the average annual precipitation is less than 400mm; unstable irrigation zone where the average annual precipitation is greater than 400mm but less than 1000mm; and supplemental irrigation zone where the average annual precipitation is greater than 1000mm.

The conditions for agricultural production are not very satisfactory in most parts of China, especially the threats from drought and water shortage are dominant. There would be no stable agriculture without irrigation in regions where the annual precipitation is less than 400mm; irrigation plays a vital role in guaranteeing good harvests in semi-humid and semi-arid regions; seasonal drought turn out to be greater harmful to agriculture in southeastern coastal areas, as a result irrigation is also a prerequisite to guarantee agricultural production.

According to the investigation, the per unit area yield of grain is approximately 6000kg per hectare in areas with irrigation facilities, while it is only around 2100kg per hectare in areas without irrigation facilities. The yield of grain from irrigated farm is 2-4 times of that from non-irrigated one. Moreover, there will be more yield increase in areas that is drier. According to the analysis, 70% of the national total grain production, 80% of the national total cotton production and more than 90% of the national total vegetable production come from arable land with irrigation facilities. Apparently, irrigation is a key and major measure to enhance the capacities to cope with drought disasters as well as to guarantee good harvests.

1 Water and Agriculture 5

Irrigation Index in Three Irrigation Zones

Irrigation zone category	Region	Crop	Dry year			Wet year		
			Total water demand (mm)	Irrigation demand (mm)	Irrigation index	Total water demand (mm)	Irrigation demand (mm)	Irrigation index
Perennial irrigation zone	Northwestern inland and middle and upper reaches of the Yellow River	spring wheat	450–520	300–450	0.7–0.9	300–450	200–350	0.7–0.8
		maize	375–450	250–350	0.7–0.8	375–450	250–300	0.7–0.8
		cotton	600–750	450–500	0.6–0.7	600–750	300–450	0.5–0.6
Unstable irrigation zone	Middle and lower reaches of the Yellow River, Huai River basin, Hai River Basin	paddy	1000–1200	600–800	0.6–0.7	850–1000	400–600	0.5–0.6
		winter wheat	600–750	300–450	0.5–0.6	500–600	200–300	0.4–0.5
		maize	450–600	300–450	0.7–0.8	300–500	100–200	0.3–0.4
		cotton	750–900	300–450	0.4–0.5	550–675	100–200	0.2–0.3
		paddy	900–1100	500–700	0.5–0.6	800–1000	300–500	0.4–0.5
	The northeast	spring wheat	300–450	80–150	0.2–0.3	225–375	0	0
		maize	400–500	100–150	0.2–0.3	300–400	0	0
Supplemental irrigation zone	Middle and lower of the Yangtze River	early rice	675–825	300–450	0.4–0.5	450–600	100–150	0.3–0.4
		late rice	825–1000	450–600	0.5–0.6	750–900	150–300	0.2–0.3
		winter wheat	400–600	50–100	0.1–0.2	225–375	0	0
		cotton	750–975	150–300	0.2–0.3	575–700	0–100	0–0.1
	The Pearl River and Minjiang River Basin, southwest	early rice	600–750	300–450	0.5–0.6	450–600	100–150	0.2–0.3
		late rice	750–825	300–450	0.4–0.5	600–750	150–300	0.3–0.4
		winter wheat	400–600	0–50	0–0.1	250–350	0	0

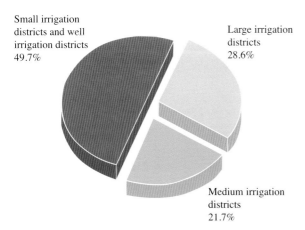

Proportion Chart of all Kinds of Irrigation Districts

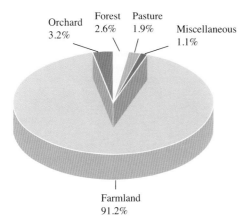

Proportion Chart of Irrigation Area of Various Crops

1.2.3 Irrigation District and Irrigation Area

In 2008, China had 6414 large and medium irrigation districts whose irrigation area is more than 10 hundreds mu, including 447 large irrigation districts. China also had numerous small irrigation districts and well irrigation districts. The command irrigation area of large irrigation districts accounts for 28.6% of the national total; the medium irrigation districts makes up about 21.7% of the national total; and the small irrigation districts, well irrigation districts and irrigation lands account for about 49.7% of the national total.

By the end of 2008, the total irrigation area in China was 64.12 million hectares, including 58.472 million hectares (command) irrigation area of farmland, 2.065 million hectares irrigation area of orchard; 1.649 million hectares irrigation area of forest, 1.214 million hectares irrigation areas of pasture; and 0.72 million hectares of irrigation area of miscellaneous.

Statistics of Command Irrigation area, Irrigation Water Use and Grain Production in Some Years

Year	Command irrigation area ($10^4 hm^2$)	Irrigation water use ($10^8 m^3$)	Percentage to total water use (%)	Irrigation area per capita (hm^2/capita)	Total grain production (10^8 kg)
1949	1600	956	92.0	0.029	1132
1957	2500	1853	90.0	0.039	1950
1965	3207	2350	85.0	0.044	1945
1980	4887	3574	80.5	0.049	3205
1988	4793	3874	78.0	0.044	3941
1993	4973	3440	66.5	0.042	4565
1998	5313	3495	64.3	0.043	5123
2000	5467	3600	64.3	0.043	4622
2006	5708	3662	63.2	0.043	4975
2007	5778	3602	61.9	0.044	5015

Notes: the command irrigation area refers to the area that can be normally irrigated under the current engineering and water source conditions in an irrigation district.

1.2.4 Irrigation Schemes

China's irrigation schemes can be divided into three big categories, i.e. water storage, water lift and water division.

(1) Water storage irrigation scheme consists of more than 86 thousand reservoirs (the storage capacity is more than 1×10^5 m^3) with a total capacity of 6.924×10^{11}m^3. approximately 6 million ponds whose water storage capacity is between 300m^3 and 100000m^3; and around 5 million pools (cellars) whose water storage capacity is less than 300m^3.

(2) Water lift irrigation scheme includes 444 thousand fixed electromechanical irrigation and drainage pumping stations with 23.95 million kW of possessing capacity; 4.388 million corollary driven wells with 40.89 million kW of possessing capacity, out of which 93% are located in the north; as well as many mobile irrigation and drainage mechanical equipments with 21.84 million kW of possessing capacity.

(3) With regard to water diversion scheme, there are numerous irrigation facilities diverting water from rivers through gravity or through building dams.

1.2.5 The Demand for Irrigation Water Conservation

According to the relevant planning formulated by the Ministry of Water Resources, P.R China, to guarantee the national food security, appropriately 46 billion m^3 of annual agricultural water saving capacity should be developed under conditions that the irrigation reliability is maintained at a certain level while keeping the totoal irrigation water use unchanged by 2030. Therefore, it is imperative to extend water saving irrigation technology on irrigated farms and to improve irrigation water management.

2 Overview of Water Saving Irrigation Development

2.1 Progress of Water Saving Irrigation Development

2.1.1 Water Saving Irrigation in Ancient and Recent Times

Ancient China had already applied the water saving technology into agricultural production, such as land leveling, narrowing furrow and basin size for irrigation, and deep plowing to loosen the soil. In the early 1930s, Pangshan Irrigation Experimental Site was established in Wujiang County of Jiangsu Province and Fenghuai District Experimental Site was established in Linhuaigang of Anhui Province to conduct scientific experiment on saving irrigation water use on farm lands.

2.1.2 Water Saving Irrigation from 1950s to 1980s

China's water saving irrigation in this period mainly focused on research and extension of individual technology. Planned water use, canal seepage control and improved furrow and basin irrigation technology et al centering on increasing irrigation water use efficiency were conducted; hundreds of irrigation experimental stations were established and the irrigation scheduling of some major crops was brought forward; "seeping paddy field with new methods and irrigating paddy field with shallow water" was extended in the south; improved furrow and basin irrigation technology was advanced in the north. Land leveling and constructing horizontal terraced field were popularized throughout the country to realize the overall planning and comprehensive management of hills, water, farmlands, forests and roads. In the 1970s, internationally advanced sprinkler irrigation and micro irrigation technologies were imported and digested into China while gradually extending canal seep control, and some cheap mechanical and automatic irrigation equipments were developed in China independently. Low pressure pipeline irrigation was stressed and extended in the 1980s.

2.1.3 Water Saving Irrigation since the 1990s

China's water saving irrigation has entered into a rapid development stage since the 1990s. "To make great efforts to popularize water saving irrigation technology" had been proposed by the government where it is requested "to promote water saving irrigation as a revolutionary measure." The governments at various levels had strengthened the organization and guidance for water saving irrigation and offered specific discounted loans and financial assistance funds to intensify technical guidance and services. In addition to this, a lot of water saving and yield increasing key counties and thousands of water saving irrigation technology demonstration plots had been established, and the main works of all of the large irrigation districts and parts of the medium irrigation districts had been rehabilitated.

By the end of 2008, 24.435 hectares of water saving irrigation schemes had been rehabilitated in accordance with the current national standards, accounting for 41.8% of the total irrigation area of farmlands.

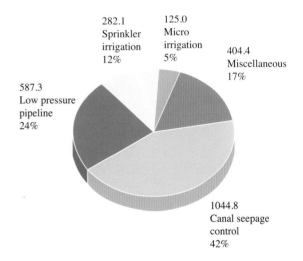

Ratio and Area of Various Saving Irrigation (unit: $10^4 hm^2$)

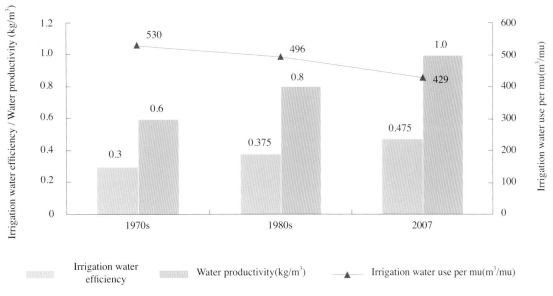

Schematic Diagram of Water Saving Efficiency

Water Saving Irrigation Area of All Provinces (Autonomous Regions / Municipalities) by the End of 2008

unit: 10^4 hm^2

Provinces (Autonomous regions / Municipalities)	Command irrigation area	Water saving irrigation area					
		Subtotal	Canal seepage control	Low pressure pipeline	Sprinkler irrigation	Micro irrigation	Miscellaneous
Total	5847.1	2443.5	1044.8	587.3	282.1	125.0	404.4
Beijing	24.1	28.7	4.4	15.7	7.2	1.4	0.0
Tianjin	34.8	22.9	8.4	13.8	0.5	0.2	0.0
Hebei	455.9	254.4	30.3	170.3	29.2	2.3	22.4
Shanxi	125.5	80.5	17.3	45.8	14.4	2.9	0.1
Inner Mongolia	287.1	200.9	68.3	86.4	44.9	1.1	0.2
Liaoning	149.3	43.4	10.3	9.9	17.2	3.0	2.9
Jilin	165.4	23.5	1.8	0.0	20.7	0.0	0.9
Heilongjiang	312.3	199.6	6.0	1.0	76.3	2.4	114.0
Shanghai	23.5	14.3	4.8	9.2	0.2	0.0	0.0
Jiangsu	381.7	153.7	101.7	6.5	1.6	0.4	43.6
Zhejiang	143.6	97.3	75.2	7.0	2.3	1.5	11.3
Anhui	345.4	76.5	18.8	6.7	7.6	0.7	42.7
Fujian	95.5	50.5	39.5	6.2	2.9	0.6	1.3
Jiangxi	184.1	24.8	11.7	0.2	0.7	0.0	12.2
Shandong	485.7	207.6	45.0	95.9	14.5	4.2	47.9
Henan	498.9	141.2	47.1	58.9	11.1	0.8	23.3
Hubei	233.0	34.7	28.2	2.8	1.3	0.6	1.9
Hunan	270.9	31.6	26.7	0.5	2.0	0.2	2.3
Guangdong	186.3	17.6	15.5	0.8	0.9	0.2	0.3
Guangxi	152.1	68.0	39.6	0.4	0.5	0.0	27.5
Hainan	24.6	10.4	6.2	0.1	0.1	0.1	4.0
Chongqing	65.9	12.9	9.9	1.9	0.5	0.0	0.6
Sichuan	250.7	109.8	94.9	4.9	4.1	0.9	4.9
Guizhou	91.7	38.0	28.1	2.0	0.4	0.8	6.8
Yunnan	153.7	49.0	39.0	4.0	0.6	0.2	5.2
Tibet	22.1	3.3	3.2	0.1	0.0	0.0	0.0
Shaanxi	130.1	84.1	48.1	21.2	3.9	1.4	9.6
Gansu	125.5	79.6	55.3	8.8	4.5	2.8	8.1
Qinghai	25.1	7.2	6.7	0.0	0.2	0.0	0.4
Ningxia	45.2	22.7	17.7	1.7	0.5	0.6	2.1
Xinjiang	357.3	254.8	135.4	4.8	11.2	95.7	7.7

2.2 Scientific and Technical Support to the Development of Water Saving Irrigation

Since the 1970s, China has attached great importance to and strengthened the research, test and extension of water saving irrigation technology suitable to the conditions of China while introduced and digested advanced water saving irrigation technologies overseas, has conducted researches of many projects in succession, such as "Research and Extension of Sprinkler Irrigation and Micro Irrigation Technologies", "Development and Extension of the Irrigation Technology Delivering Water with Low Pressure Pipelines", "Research and Extension of Water Saving Type Agricultural Technical System in North China", "Research and Demonstration of the National Water Saving Agricultural Technology", "Demonstration Projects of the Scientific and Technical Industry of Efficient Agricultural Water Use", "Research and Development of the Technical System and New Products of Modern Water Saving Agriculture", "Research on the Key Support Technology of Rehabilitation Projects in Large Irrigation District", as well as "Research and Application of Rainwater and Flood Water Harvesting Technology", and has made great achievements in the research on water saving irrigation technologies.

List of Research Results from Water Saving Irrigation Technology

Category	Technical research results
Rational water allocation and efficient water use technologies	Conjunctive surface water and groundwater exploitation and use technology in irrigation district; "Reservoir groups strung together by canals" type efficient water use technology in surface water irrigation district; The efficient water use technology of irrigation return water, brackish water, and treated industrial and urban sewage water in agricultural irrigation; The efficient water use technology of rain water harvesting in agricultural irrigation; Irrigation water monitoring and information collection and processing technology in irrigation district
Water conveyance and allocation technology in irrigation district	Canal freezing resistance and seepage control water conveyance technology; Rehabilitation technology of canal structures; Irrigation technology delivered water with low pressure pipelines; Automatic water allocation control technology of water conveyance and water allocation canals/pipelines; Water measurement technology of canals/pipelines
Water saving irrigation technology on farmland	Sprinkler irrigation and micro irrigation technology; Drip irrigation technology under plastic film; Gated-pipe irrigation technology on farmland; Alternate surge irrigation technology; Laser controlled land leveling technology; Improved furrow and basin irrigation technology
Agronomic and biological water saving technology	Water saving and efficient irrigation scheduling of major crops; Farmland cover and improved tillage water saving technology; Water and fertilizer coupling and blended fertilization technology; Chemicals control evapotranspiration and soil moisture; retention technology; Breeding and cultivating technology of anti-drought varieties

2.3 Water Saving Irrigation Facilities and Equipments

China has initially established the production system of conventional water saving irrigation facilities and equipments, and researched, developed and produced various types of water saving irrigation equipments suitable to the conditions of China.

Types and names of various water saving equipments

Euipment type	Name
Sprinkler irrigation equipment	Various sprinkler heads and tapes; Thin metal pipeline system that can be dismounted quickly; Medium and small sprinkling set; Capstan and central pivot sprinkling set etc.
Micro irrigation equipment	Various micro sprayers; Various drip emitters and drip lines; Filter, fertilizer tank, pipe and pipe fitting etc.
Surface irrigation equipment	Various pipe and pipe fitting of low pressure water conveyance; On-farm gated pipeline; Hydrant; Alternate surge irrigation equipment etc.
Water measurement equipment	Various water measurement meters and equipments on canals and pipelines

3 Strategy on Water Saving Irrigation Development

3.1 General Train of Thought

China's water saving irrigation focuses on increasing water use efficiency and productivities, and aims to improve agricultural production conditions, to enhance overall agricultural production capacity, to increase farmers' income and to improve the ecological environment through adopting engineering, agronomic, biological and management et al comprehensive water saving measures. The areas that were given priority in the development of water saving irrigation in China are those water shortage areas, ecologically and environmentally vulnerable areas, water abundant areas but with serious non-point source pollution, as well as suburbs of water deficient cities et al.

3.2 The Way Forward toward Water Saving Irrigation by Regions

3.2.1 Northeastern Region

The Northeastern region mainly include the whole areas of Heilongjiang Province, Liaoning Province and Jilin Province and the four leagues in the east of Inner Mongolia Autonomous Region. It is the main production base of maize, paddy rice, wheat and soybean in China.

The Northeastern region is characterized by serious spring drought with long duration.

It is planning to implement rehabilitation of irrigation districts in this region; to cut down the planting area of paddy rice and to popularize controlled irrigation technology applied in paddy in water scare areas; to rationally control the water use in salt-leaching and alkali control in low-lying waterlogging-prone areas and alkaline areas; to vigorously extend sprinkler irrigation in centralized and linked-up together planting areas; to build water saving irrigation forage base in deteriorated and eroded grassland areas and to carry out rotation grazing or grazing prohibition; to reserve enough ecological water for the protection of wetlands.

3.2.2 Huang-Huai-Hai Region

The Huang-Huai-Hai region mainly includes the whole areas of Beijing Municipality, Tianjin Municipality, Hebei Province, Shandong Province, Anhui Province and parts of Shanxi Province, Inner Mongolia Autonomous Region and Henan Province. They are the main production areas of wheat, maize, and cotton et al agricultural products in China.

Huang-Huai-Hai region is a serious water shortage area where the groundwater has been over-exploited, parts of the coastal areas have been intruded by sea water, many rivers have been cut off, and the ecological environment is deteriorated.

It is planning to implement rehabilitation of irrigation districts in this region; to rigorously control the development scale of irrigation districts in water over-exploitation areas, to practice integrated well and canal irrigation, and to carry out conjunctive surface water and groundwater allocation; to develop and apply brackish water et al low quality water in well irrigation districts, to rationally control groundwater exploitation, and to recharge groundwater with flood water in rainy seasons to supply water sources in areas where conditions permit; to extend canal seepage control and low pressure pipeline water conveyance technologies as well as improved surface irrigation technology; to gradually adjust the cropping pattern and to properly reduce the planting area of winter wheat et al high water-consuming crops in Haihe River Basin. There are many large and medium cities in this region, so sprinkler irrigation and micro irrigation technologies should be popularized in the planting area of vegetables et al cash crops.

3.2.3 The Middle and Upper Reaches of the Yellow River

The middle and upper reaches of the Yellow River refers to the area of the Yellow River Basin above Sanmenxia Gorge. It includes the whole area of Shaanxi Province, Gansu Province, Ningxia Hui Autonomous Region and parts of Shanxi Province, Inner Mongolia Autonomous Region, Henan Province and Qinghai Province. The source of irrigation water in this region is mainly from the Yellow River and its tributaries.

It is planning to strictly diver water in accordance with the amount of water allocated to the irrigation districts diverting water from the Yellow River in Ningxia and Inner Mongolia Autonomous Region; to rationally adjust the layout of canal systems, and to line the mainstay canals to prevent seepage; to strengthen land leveling, to improve furrow and basin irrigation technology, and to extend farm cover technology; to conduct integrated well and canal irrigation in areas where conditions permit, to rationally control the depth to groundwater table, and to prevent secondary soil salinization; to strictly control the total exploitation of groundwater in confined aquifers, and to recharge groundwater through diverting flood water stored in rainy seasons in areas where groundwater has been over-exploited; to strictly control the stock capacity in areas where the grassland has been deteriorated and eroded, to build water saving irrigation forage base, and to return grazing to pasture integrated with animal husbandry measures to protect the ecology and to increase the farmers' and herders' income.

3.2.4 Inland Region

The Inland region includes Xinjiang Uygur Autonomous Region, Gansu province, Inner Mongolia Autonomous Region, and parts of Qinghai province with scare precipitation and strong evaporation. There would be no agriculture without irrigation.

It is planning to implement rehabilitation of irrigation districts in this region; to extend technologies of seepage prevention of canals and water conveyance through pipelines, to level land, to cut down land size, and to improve furrow and basin irrigation technology; to strictly diver water in accordance with the amount of water allocated, to control the scale of irrigation districts, and to reserve enough water for the ecological environment while popularizing water conservation in Tarim River Basin, Heihe River Basin and Shiyanghe River Basin where the ecological environments are vulnerable; to extend the technology of drip irrigation under plastic film in the production unit of cotton and tomato et al cash crops; to extend technologies of rainwater harvesting for irrigation, protective tillage, farm cover, and biological and chemical drought control et al.

3.2.5 The Yangtze River Region

The Yangtze River region mainly consists of the whole area of Shanghai Municipality, Sichuan Province, Chongqing Municipality, Hubei Province, Hunan Province and Jiangxi Province and parts of Henan Province, Shaanxi Province, Anhui Province and Jiangsu Province. They are the important production units of paddy and rapeseed.

There is abundant precipitation, widely distributed lakes and rich water resources in the Yangtze River region.

It is planning to implement rehabilitation of irrigation districts centering on the requests for modern agriculture in this region; to extend canal lining and controlled paddy irrigation technologies; to adopt sprinkler irrigation and micro irrigation technologies according to local conditions in the planting area of cash crops; to construct small-scale and micro-scale water storage schemes, to exploit water sources to resist drought, and to extend advanced and practical water saving irrigation technology in hilly areas; to accelerate the pace in the construction of garden tillage in strict accordance with the requests for efficient water use and modern agriculture in economically advanced coastal areas.

3.2.6 The Pearl River Region

The Pearl River region includes the whole area of Guangdong Province and Fujian Province and parts of Guangxi Zhuang Autonomous Region where spring and autumn droughts occur frequently and there is relatively serious water pollution.

It is planning to implement rehabilitation of irrigation districts centering on the requests for modern agriculture in this region; to extend canal lining and controlled paddy irrigation technologies; to construct small-scale and micro-scale water storage schemes, to exploit water sources to resist drought in hilly areas; and to adopt sprinkler irrigation and micro irrigation technologies according to local conditions in the planting area of cash crops.

3.2.7 The Southwestern Rivers Region

The Southwestern Rivers region mainly consists of the whole area of Yunnan Province and Guizhou Province and parts of Guangxi Zhuang Autonomous Region where agriculture is weak at resisting natural disasters and the irrigation and drainage infrastructures lag behind.

It is planning to implement rehabilitation of irrigation districts in this region; to extend canal lining and controlled paddy irrigation technologies; to construct small-scale and micro-scale water storage schemes, to exploit water sources to resist drought in hilly areas; and to adopt sprinkler irrigation and micro irrigation technologies according to local conditions in the planting area of cash crops.

4 Technical System and Index of Water Saving Irrigation

4.1 Technical System of Water Saving Irrigation

The technical system of water saving irrigation consists of rational water resources exploitation and use, water saving irrigation engineering technology, irrigation water use management technology, agronomic water saving technology, and biological water saving technology.

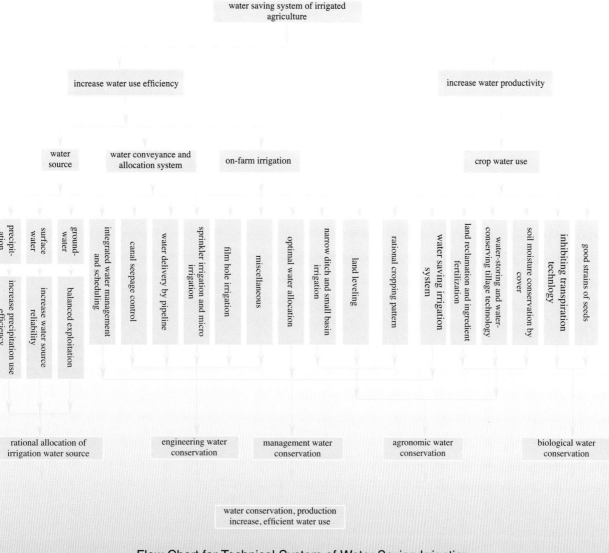

Flow Chart for Technical System of Water Saving Irrigation

4.2 Technical Index of Water Saving Irrigation

The main technical indexes of water saving irrigation are irrigation water use efficiency and crop water productivity.

4.2.1 Irrigation Water Use Efficiency

Irrigation water use efficiency can be evaluated by canal system water use coefficient, on-farm water use coefficient and irrigation water use coefficient. The current national standard "Technical Specifications on Water Saving Irrigation Schemes" (GB/T 50363-2006) has made relevant regulations on the above coefficients.

Specifications on Irrigation Water Use Coefficients

Item		Coefficient		Remarks
Canal system water use coefficient	Surface water irrigation district	Large-sacle irrigation district	≥0.55	The coefficient can be reduced by 0.10 within the above range in irrigation districts where well irrigation and canal irrigation have been fully integrated; the coefficient can be reduced according to the ratio of the integrated well and canal irrigation area to total irrigation area in irrigation districts where well irrigation and canal irrigation have been partially integrated
		Medium irrigation district	≥0.65	
		Small-scale irrigation district	≥0.75	
	Well irrigation district	Canal lining	≥0.9	
		Water delivery with pipeline	≥0.95	
On-farm water use coefficient	Paddy irrigation district		≥0.95	
	Dry farming irrigation district		≥0.90	
Irrigation water use coefficient	Large-scale irrigation district		≥0.50	The coefficient in integrated well and canal irrigation district can be determined on the weighted average of the total water use by canals and wells
	Medium irrigation district		≥0.60	
	Small-scale irrigation district		≥0.70	
	Well irrigation district		≥0.80	
	Sprinkler irrigation district		≥0.80	
	Micro irrigation district		≥0.85	
	Drip irrigation district		≥0.90	

4.2.2 Crop Water Productivity

It is stipulated by China's current national standard "Technical Specifications on Water Saving Irrigation Schemes" (GB/T 50363-2006) that the overall production capacity (total output of grains and cotton) should be increased by more than 15% and the water productivity of grain crops should be increased by more than 20% and should be no less than 1.2kg/m^3 after implementation of water saving irrigation.

5 Main Measures of Water Saving Irrigation

5.1 To Rationally Allocate and Efficiently Use Irrigation Water Sources

Adjusting the temporal and spatial distribution of various water sources in river basins and regions through structural and non-structural measures and conducting conjunctive water allocation among multiple water sources so that to achieve the maximum irrigation benefits.

5.1.1 Surface Water Irrigation District

Developing "reservoir groups strung together by canals" type irrigation system to realize conjunctive water allocation among large, medium, small, water storage, water diversion and water lift schemes; recharging groundwater with irrigation water return as well as foreign water diverted by canals or rainwater harvested in drainage ditches, and developing integrated well and canal irrigation.

5.1.2 Well Irrigation District

Reducing groundwater exploitation and rationally exploiting and using brackish water to realize irrigation directly using brackish water or using mixed fresh water and brackish water.

5.1.3 Rainwater Harvesting and Use

In arid hilly regions, rainwater can be harvested by various means, such as slope surface and roads, which can harvest rainfall efficiently, and then be diverted and stored into cellars and pools as crucial water for crops to fight against drought.

5.1.4 Sewage and Waste Water Use

Wastewater discharged from urban domestic uses or industries should be treated to a standard that meet irrigation water quality. This kind of water can be used to irrigate non-immediate-edible crops.

5.2 To Increase Water Conveyance and Allocation Efficiency

5.2.1 Canal Lining

Canal lining are mainly used in surface water irrigation districts. Earth materials, aggregrated rock, coating materials and concrete or asphalt concrete et al materials can be used to line canals to form impervious layer so that to increase irrigation water conveyance and allocation efficiency.

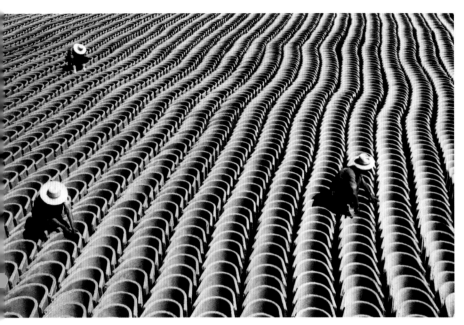

5.2.2 Water Delivery with Low Pressure Pipelines

Water delivery with low pressure pipeline system is mainly applied in well irrigation districts. Pipes such as embedded hard plastic pipes, embedded concrete pipes, ground nylon-coated or vinylon-coated hoses, or ground plastic membrane hoses can be used to deliver water to increase the irrigation water conveyance and allocation efficiency.

5.2.3 Rehabilitation of Canal Systems and Structures of Pumping Stations

In order to meet the requirements of irrigation system, to facilitate the general public's production, living and transportation as well as to facilitate water allocation and measurement, there is a need to rehabilitate the gates, culverts, bridges, aqueducts and hydraulic drops et al canal system structures so that to increase the irrigation water conveyance and allocation efficiency; there is also a need to test the efficiency and to rehabilitate the devices of water-lifting pumping stations.

5.3 To Increase Water Use Efficiency of Surface Water Irrigation

The main measures to increase the water use efficiency of surface water irrigation include land leveling, improved furrow and basin irrigation, film hole irrigation, intermittent irrigation, gated-pipe irrigation and controlled irrigation of paddy.

5 Main Measures of Water Saving Irrigation 39

5.4 Sprinkler Irrigation

The sprinkler irrigation types in common use include sprinkler irrigation system in pipe network, small mobile sprinkling set, and large and medium sprinklers. Priority should be given to gravity sprinkler irrigation in hilly areas with high water and low land.

5.5 Micro Irrigation

Micro irrigation is mainly applied in the irrigation of vegetables and fruit trees et al cash crops. Priority should be given to gravity micro irrigation in hilly areas with high water and low land.

5.6 Dibble Seeding and Water Injecting Technology

Hole irrigation for seedling has been applied in over 4.667 million hectares of arable land in Northeastern region each year. Travelling water-injecting hill drop planter can be applied at the sowing and seedling stages to finish trenching, water injection, dibble seeding, fertilization, earthing once so that to meet the water demand for seed sprouting and to keep a full stand of seedlings.

5.7 Agronomic Water Saving Technologies and Measures

Agronomic water conservation is not only an indispensable component of the technical system of water saving irrigation, but also a main measure for efficient use in rainfed agriculture. The measures mainly include: to apply drought-resistant variety; to apply soil moisture conservation technology by tillage and land cover, to apply water-fertilizer coupling technology, and to apply chemical agent. Out of which, the coupling model and technology focusing on fertilizer, water and crop yield can increase the fertilizer use efficiency by 3%-5% and increase the yield by 20%-30% under conditions of not increasing the amount of use of fertilizer and water.

The measures also include rationally applying water retention agent, compound coating agent, fulvic acid, multifunctional evaporation control and drought resistant agent and "ABT" rooting power et al to control over-evapotranspiration and extravagant water consumption during crop growth period and to promote the use of water stored in deep soil by crop root system so that to alleviate the impact from drought disasters.

5.8 Management Water Saving Technologies and Measures

The water saving technologies and measures in the aspect of management mainly include: total amount control and quota management in irrigation water use, water saving irrigation scheduling, soil moisture monitoring and irrigation forecasting technology, water measurement technology in irrigation district, and automatic monitoring and control technology in irrigation district et al.

5.8.1 Total Amount Control and Quota Management in Irrigation Water Use

The total amount of irrigation water use should be controlled according to the carrying capacity of water sources in an irrigation district; water users should apply water scientifically based on the assigned water use index and the water saving irrigation quota; water fees should be charged on a volumetric basis in an irrigation district and extra fee should be charged if extra water is used.

5.8.2 Water Saving Irrigation Scheduling

Deficit irrigation, drought resistant irrigation and low quota irrigation should be adopted in arid and semi-arid areas to rationally reduce irrigation times, to reduce irrigation quota and to irrigate at critical time. Meanwhile, restraining the growth of seedling, forcing and retarding culture et al technologies should be adopted to reduce on-farm evapotranspiration; "thin-shallow-wet-dry" et al water saving irrigation system should be extended in the planting area of paddy to significantly reduce irrigation water use.

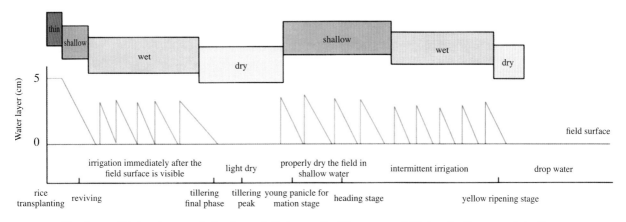

On-farm Water Layer Adjustment Schematic Diagram of "thin-shallow-wed-dry" Irrigation Method for Paddy Rice

5.8.3 Soil Moisture Monitoring and Irrigation Forecasting Technology

Tensiometer, neutron method and electric-resistivity method et al technical measures in coordination with weather forecast should be applied to monitor soil moisture and to forecast on proper irrigation time and irrigation volume.

5.8.4 Water Allocation Technology in Irrigation District

Scheme on optimal water allocation in an irrigation district should be compiled on the basis of the available irrigation water supply, the technical parameters of various water delivery canals, crop water requirement and water production function and by system engineering method in an attempt to minimize the water loss in conveyance and allocation as well as to maximize the production.

5.8.5 Water Measurement Technology in Irrigation District

The water measurement techniques such as gaging flow with open channels, ultrasonic and electromagnetism, can be used to optimally allocate, measure and charge water so that to enhance the farmers' consciousness in water conservation and to increase the water use efficiency of an irrigation district.

5.8.6 Automatic Water Monitoring Technology in Irrigation District

With the application of automatic monitoring instruments and meters, the parameters like canal water level, discharge, sediments concentration, soil moisture and pump operation situation et al can be collected and analyzed on a real time basis; then the optimal plan can be selected based on a ready-made software, and wired or wireless transmission mode can be used to control the numbers of pumps in operation and the openning of gates to realize scientific irrigation and to increase irrigation water use efficiency in coordination with the on-farm water saving techniques.

6 Laws, Regulations and Policies on Water Saving Irrigation

6.1 Laws and Regulations

6.1.1 《Water Law of the People's Republic of China》

It is stipulated by the "Water Law of the People's Republic of China", amended and put into effect on Oct. 1, 2002, that "the State strictly encourages water conservation, vigorously popularizes water saving measures, extends state-of-the-art water saving technologies, develops water saving industries, agriculture and services, and builds a water saving society". "the governments at various levels should popularize water saving irrigation and water saving technologies, and take essential measures to prevent seepage in agricultural water storage and water delivery schemes to increase agricultural water use efficiency".

6.1.2 《Agriculture Law of the People's Republic of China》

It is stipulated by the "Agriculture Law of the People's Republic of China", amended and put into effect on March 1, 2003 that "the governments and agricultural production and management organizations at various levels should strictly control the occupation of irrigation water sources by non-agricultural development according to law, prohibit any organization and individual illegally occupy or damage the irrigation and drainage infrastructures. The state will offer significant support to the development of water saving agriculture in water scare areas".

6.1.3 《Grassland Law of the People's Republic of China》

It is stipulated by the "Grassland Law of the People's Republic of China", amended and put into effect on March 1, 2003, that "the local governments above county level should support the construction of hydraulic facilities and develop water saving irrigation in grassland areas to improve the people and livestock access conditions to drinking water".

6.2 Policies for Agricultural Water Conservation

The state and local governments have launched a number of policies to encourage the development of agricultural water conservation.

6.2.1 《General Outline on Policies for Water Conservation Technology in China》

The State Development and Reform Commission and the Ministry of Science and Technology together with the Ministry of Water Resources, the Ministry of Construction and the Ministry of Agriculture enacted the "General Outline on Policies for Water Conservation Technology in China" in 2005, out of which it is posed that "agriculture is still the primary water user, and developing efficient and water saving agriculture is a fundamental strategy of the country".

6.2.2 《China's Agenda in the 21st Century - White Paper on China's Population, Environment, and Development in the 21st Century》

It is advanced from the "China's Agenda in the 21st Century - White Paper on China's Population, Environment, and Development in the 21st Century" published by the State Council in 1994, that "China should strengthen the construction of irrigation and drainage schemes and their auxiliary projects, expand irrigation area, develop water conservation technology, and increase irrigation water use efficiency".

6.2.3 Water Permitting and Water Pricing Policies

The State Council and relevant departments have launched many systems and methods such as "Regulations on Water Permitting and Water Charging Management", "Notice on Advancing Water Price Reform, Promoting Water Conservation and Protecting Water Resources", "Regulations on Water Pricing Management of Hydraulic Schemes", "Tentative Measures on Water Allocation", "Regulations on Water Permitting Management" et al in which many policies to promote water conservation have been raised.

6.2.4 Governments at Various Levels Increase Financial Support to Water Conservation in Agriculture

In the past several decades, the governments at various levels have taken measures to support the development of water conservation schemes in agriculture by providing guidance in policies and by providing financial support, have gradually increased the support strength, and have arranged specific funds in rehabilitation of large and medium irrigation districts, in water saving and yield increasing key counties, in water saving demonstration plots, in construction of small irrigation and drainage schemes, and in agricultural water conservation activities carried out by rural communities and farmers.

6.2.5 Water User Participation

Various kinds of water user participations with water user association as the major type have been advocated across China to accelerate the popularization of agricultural water conservation technology. Currently approximately 40 thousand water user associations have been established in China and participated in the management of 6.667 million hectares of irrigation area.

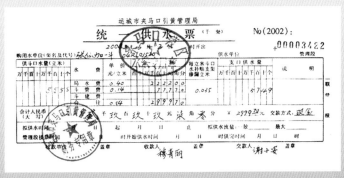

6.3 Technical Standards on Water Saving Irrigation

A series of technical standards on water saving irrigation have been enacted in China, which include "Technical Regulations on Water Saving Irrigation Schemes" (GB/T 50363-2006), "Technical Regulations on Sprinkler Irrigation Scheme" (GB/T 50085-2007), "Technical Regulations on Micro Irrigation Schemes" (GB/T 50485-2009), "Technical Regulations on Canal Seepage Prevention Schemes" (SL 18-2004), and "Technical Regulations on on-farm Irrigation Schemes Delivered Water by Low Pressure Pipelines" (GB/T 20203-2006) et al in terms of project construction and management; and "Technical Regulations on Application of Rolling Sprinkling Set" (SL 95-2004) and "Rotating Sprinkler Heads" et al in terms of water saving equipments and facilities.

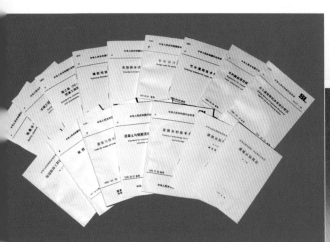

7 Effects from Water Saving Irrigation

7.1 Increased Irrigation Water Use Efficiency

The average irrigation water use efficiency in China has increased from about 30% estimated in the 1970s to 47.5% in 2007, while the per unit area irrigation water use has dropped from 530m^3/mu to 429m^3/mu. The command irrigation area has increased from 48.87 million hectares at the end of 1970 to 57.782 million hectares in 2007 under conditions that the total irrigation water use not increased on the whole.

Taking the projects of rehabilitation of large irrigation districts as an example, China invested a total of RMB 21.26 billion Yuan to rehabilitate the ill, dangerous, and throat-seizing works and severely leaky canals in 363 irrigation districts, and rehabilitated, improved and newly increased a total of 6.667 million hectares of irrigation area during the ten years between 1996 and 2006. According to the statistics, the canal water use coefficient had been increased from 0.49 to 0.54, the irrigation water use efficiency had been increased from 42.1% to 47.8%, the average irrigation water use per mu had been dropped from 529m^3 to 469m^3, and the water saving potentials had been newly increased by 12.5 billion m^3 after the implementation of these projects.

7.2 Increased Water Productivity and Overall Agricultural Production Capacity

Based on the estimation and analysis of 363 large irrigation districts which have been rehabilitated, the average yield of grain crops per mu has been increased by 30kg/mu, and the total grain production has been increased by 11.5 billion kg; the double benefits that the grain production capacity has been newly increased by 1kg and the water saving potentials has been newly increased by 1.1m^3 with 1.84 Yuan per mu of national investment have been gained.

7.3 Speeded up the Transformation from Traditional Agriculture to Modern Agriculture

The development of water saving irrigation has accelerated the process of China's agriculture transformation from a traditional one to a modern one. The dynamic integration of good irrigation and drainage infrastructures, advanced agricultural technologies and improved seed strains has transformed many water saving projects from scattering management to the management like farmers cooperation organization, company plus farmer household, and big household contracting et al. This kind of transformation has promoted the regionalized cultivation, dimensional management, and industrialized production of agriculture, and has significantly increased the agricultural production and the farmers' income. In addition to this, it has also created conditions for the surplus labor shifting from the primary industry to the secondary and tertiary industries.

7.4 Relieved the Contradictions between Water Supply and Water Demand, Improved the Ecological Environment in Some Areas

After the implementation of water saving irrigation, a part of the saved irrigation water can be used to increase the irrigation water use efficiency, and to improve and to increase the irrigation area; a part can be shifted to industrial and domestic uses; and a part can be supplied for ecological uses. It is found from the mid-term project assessment report on rehabilitation of large irrigation districts compiled by China International Consulting Corporation that, comparing with 1998, in the rehabilitation projects of large irrigation districts implemented before 2004, 30.1% of the saved water was used for ecological environment and 16.1% was used for urban and industrial water supply, therefore notable social, ecological and environmental benefits had been achieved.

7.5 Promoted the Development of Water Saving Irrigation Equipment and Facility Industry

Generally speaking, the investment in water saving irrigation facilities and materials, including cement, steels, plastic pipes, sprinkling machines and micro irrigation materials accounts for two thirds of the total investment in the construction of water saving irrigation schemes. The rapid development of water saving irrigation has driven the research, development, production, and application of water saving equipments and facilities. Currently there are about 200 enterprises in China manufacturing water saving equipments and facilities and their annual values of sales has reached up to RMB 10 billion Yuan.

8 Conclusions

Although great achievements have been made in China in water saving irrigation, there still exist a lot of weaknesses, difficulties and problems. Confronting with the gradually rigorous water crisis and the request to the national food security when the population peaks, China is projected that by 2030, the ratio of water saving irrigation area to command irrigation area should reach over 80%, the irrigation water use per mu in normal year should be controlled with in 390m^3, the national average on-farm irrigation water use coefficient should be over 0.6, and 46 billion m^3 of annual agricultural water saving capacity should be developed while keeping the total irrigation water use unchanged so that to make active contributions in all-round advancing the development of water saving society.

The cause of China's water saving irrigation is facing new opportunities and challenges, and the cause of China's water saving irrigation is a long-term and arduous task.